BEE FARMING (APICULTURE) FOR BEGINNERS

Buzzing Towards Success: Unlock the Secrets of Honey and Hive Management.

BRIYAN GREENWALT

Table of Contents

CHAPTER ONE..............................11

Understanding the Fascinating World of Bees: 11

The Importance of Bees in Agriculture:............................11

Bees are classified into three types..12

Benefits of Beekeeping for Beginners: 12

Choosing the Ideal Location for Your Bee Farm 13

Legal Considerations and Zoning Regulations............................13

Importance of Water Sources for Bees..14

CHAPTER TWO..............................15

Creating a Bee-Friendly Environment 15

Selecting Suitable Hive Types .15

Exploring Bee Sources 16

Understanding Bee Packages:..17

Catching Swarms: 18

Transportation and Installation Procedures.............................18

Bee Inspection and Health Assessment: 19

Hive Inspection Fundamentals: Frequency and Procedures 20

CHAPTER THREE 21

Identifying healthy brood patterns. 21

Monitoring for pests and diseases..................................21

Managing the Bee Population..22

Seasonal Hive Maintenance 23

CHAPTER FOUR 25

Knowing When to Harvest 25

CHAPTER FIVE 29

Importance of Providing Adequate Nutrition to Bees 29

Understanding Pollination: The Importance of Bees and Crops 30

Supplemental Feeding: Sugar Syrup and Pollen Patties 31

Natural Pest and Disease Control Methods. 31

CHAPTER SIX 33

Recognizing and treating common bee ailments 33

Integrating Medicine: When and How to Use Them 34

Evaluate Hive Strength for Expansion. 35

Splitting Hives: Advantages and Procedures............................35

Purchasing Additional Colonies 36

CHAPTER SEVEN 37

Queen Rearing: Methods for Sustainable Beekeeping..............37

Managing the Transition to Commercial Beekeeping..........37

Scaling Up: Infrastructure and Business Consideration 38

Branding and Packaging: 39

Identifying Target Markets: 39

Pricing Strategies 40

CHAPTER EIGHT 41

Developing Sales Channels 41

Building Customer Relationships 41

Treating Bee Stings: First Aid and Prevention 42

Addressing Neighborhood Concerns: Education and Communication...........................43

Managing Aggressive Bees: Tips to Calm Them..........................43

Legal Issues and Regulations: Permits and Liability Insurance. 44

CHAPTER NINE 45

Managing Swarm Calls: Protocols and Safety Measures...................45

Identifying and Addressing Queen Issues: 45

Coping with Robbery Behavior: Prevention and Intervention....... 46

Managing Varroa Mites: An Integrated Pest Management Approach 47

Recognize and Prevent Colony Collapse Disorder (CCD). 48

CHAPTER NINE 49

Handling Wax Moths, Small Hive Beetles, and Other Pests............. 49

Joining Beekeeping Associations: Benefits and Opportunities. 50

Taking part in workshops, conferences, and training programs 51

Building a Beekeeping Library: Recommended Books and Publications. 52

CHAPTER TEN **53**

Mentorship and Networking: Learn from Experienced Beekeepers 53

© 2024 [Briyan Greenwalt]. Reserved all rights.

All content in this book cannot be duplicated, shared, or conveyed in any way, including photocopying, recording, or other electronic or mechanical techniques, without the author's prior consent in writing. The only exceptions are short quotes included in reviews and certain other noncommercial uses allowed by copyright laws.

Disclaimer

The author's study, experience, and understanding of livestock management constitute the basis of the material in this book. Concerning the material contained in this book, the author is neither connected to, nor has any affiliation with, any organization, corporation, or person.

The author makes every effort to ensure that the material is accurate and thorough, but any errors or omissions, as well as any results resulting from the use of this information, are not covered by this statement. We strongly advise readers to consult a specialist for advice unique to their situation.

CHAPTER ONE

Understanding the Fascinating World of Bees: Bees are fascinating creatures that build intricate societies within their colonies. They communicate using elaborate dances and make honey by gathering nectar from flowers. Understanding their behavior and biology is critical to effective beekeeping.

The Importance of Bees in Agriculture: Bees pollinate crops, resulting in fruit and seed production. Without bees, many plants would struggle to reproduce, resulting in a reduction in food supply and biodiversity.

Basic Bee Biology: Roles of the Queen, Drones, and Workers: In a beehive, the queen bee lays eggs, drones mate with the queen, and workers do a variety of jobs such as foraging, nursing larvae, and maintaining the hive construction. Knowing these roles helps beekeepers manage their colonies more successfully.

Bees are classified into three types: honeybees, bumblebees, and solitary bees. Honeybees dwell in colonies, bumblebees in smaller colonies, and solitary bees live by themselves. Each variety has distinct qualities and behaviors that influence its suitability for beekeeping.

Benefits of Beekeeping for Beginners: Beekeeping has various advantages for beginners, including a sustainable source of honey, beeswax, and other hive products, as well as the ability to contribute to pollinator conservation efforts and connect with nature.

Safety Measures: Essential Gear and Precautions: A bee suit, gloves, and a smoker are required to guard against stings and properly manage bees. Beginners should also learn proper hive inspection techniques and exercise caution when dealing with bees to avoid accidents.

Choosing the Ideal Location for Your Bee Farm

Consider your region's climate to ensure it is suited for bees year-round. Bees thrive in sunny environments, so make sure your hives receive enough sunlight. Accessibility is essential; find a location where you can readily watch and tend to your bees without interference.

Legal Considerations and Zoning Regulations

Before you build your apiary, make sure you understand and follow local beekeeping regulations and zoning restrictions. Obtain any permits or licenses required in your region. By following legal standards, you can prevent potential fines or problems with neighbors or authorities.

Importance of Water Sources for Bees

Create a nearby water supply for your bees, such as a little pond or birdbath. Bees require water for hydration, cooling the colony, and diluting honey before ingestion. Ensuring that they have easy access to clean water benefits both their health and productivity.

CHAPTER TWO

Creating a Bee-Friendly Environment

Plant a variety of flowering plants in your beekeeping region to offer bees plenty of nectar and pollen. Choose plants that bloom at different times to offer a steady supply of food throughout the year. Avoid using pesticides and chemicals that may harm bees and their habitats.

Selecting Suitable Hive Types: Langstroth, Top Bar, and Warre

Choose a hive type that best fits your beekeeping aims and interests. Langstroth hives are ubiquitous and easy to handle because of their removable frames. Top bar hives are simpler and more natural, making them perfect for new beekeepers who value sustainability. Warre hives are similar to top bar hives, however they offer superior insulation and require fewer inspections. Choose the hive design that best suits your beekeeping philosophy and resources.

Exploring Bee Sources: Nucleus Colonies, Packages, and Swarms When starting beekeeping, beginners can obtain bees from a variety of sources. Nucleus colonies, or small colonies with a queen and a few thousand worker bees, are ideal for beginners because they are already established. Another alternative is to purchase bee packages, which include worker bees and a queen and are frequently supplied in screened boxes. Swarms, or natural colonies created when a queen leaves with a swarm of worker bees, can also be captured and hived.

Understanding Bee Packages: What to Look For Bee packages are a popular choice among novices due to their low cost and simplicity of transportation. Beginners should choose bee packages that have a healthy queen and an adequate number of worker bees. Before installation, make a thorough inspection of the package for symptoms of sickness or stress.

Evaluating Nucleus Colonies: Benefits and Considerations Nucleus colonies, or "nucs," provide newbies a head start in beekeeping because they already have a laying queen and a working bee population. However, beginners should carefully assess the health and strength of the nucleus colony before purchasing. To ensure a successful start in beekeeping, look for symptoms of disease, adequate brood patterns, and a healthy queen.

Catching Swarms: Safety Tips and Techniques Catching swarms can be an exciting way for beginners to get bees, but it takes caution and good technique. Before attempting to catch a swarm, make sure you have the appropriate protection gear, such as a bee suit and gloves. Approach the swarm with caution and gentleness, using a swarm box or hive to capture them without causing harm. Be aware of any adjacent obstacles and always emphasize safety.

Transportation and Installation Procedures Once bees have been acquired, proper transportation and installation are critical to their survival. To avoid overheating or stress, keep the bees safe and well-ventilated during transportation. When you arrive at the hive location, carefully put the bees into their new hive while ensuring the queen is safely introduced. To reduce interruption and stress, continuously monitor the bees during the installation process.

Bee Inspection and Health Assessment: Regular inspections are required to keep bee colonies healthy. Beginners should carefully study the hive's brood patterns, honey storage, and overall bee population. Look for indicators of disease, pests, or queen-related problems, such as a shortage of eggs or erratic brood patterns. Beginners may ensure the long-term health and success of their beekeeping endeavors by doing thorough inspections and addressing potential problems early on.

Hive Inspection Fundamentals: Frequency and Procedures

Regular hive inspections are essential for a healthy bee colony. During the active season, inspect at least once every week. Start by gently burning the hive entrance to quiet the bees. Carefully open the hive, evaluating each frame for disease, pests, and queen performance. Examine the colony's size, honey storage, and brood trends. Close the hive gently to reduce disturbance.

CHAPTER THREE

Identifying healthy brood patterns.

Healthy brood patterns are critical for a flourishing hive. During inspections, check the brood frames for consistency.

Look for a consistent pattern of capped brood cells containing larvae at different stages of development. A healthy brood pattern suggests a productive queen and a thriving colony. Irregular patterns or spotty brood can indicate problems such as illness or a failing queen.

Monitoring for pests and diseases

Beekeeping requires vigilant pest and disease monitoring. Keep a look out for common pests like Varroa mites, hive beetles, and foulbrood. Check mite levels regularly with a sticky board or alcohol wash. Look for indicators of sickness, such as discolored brood, unpleasant scents, or strange behavior. Swift intervention is essential for preventing infestations and maintaining hive health.

Managing the Bee Population: Swarming Prevention and Control

Swarming is a natural characteristic of honeybee colonies, however, it can diminish productivity if not controlled.

To prevent swarming, make sure the hive has enough space by adding supers before the brood chamber becomes congested. Inspect the hive for queen cells, which indicate swarming readiness, and split the colony if necessary. To properly manage swarming behavior, use tactics such as requeening and supplying swarm traps.

Seasonal Hive Maintenance
Spring, Summer, Autumn, and Winter

Every season brings new problems and opportunities for beekeepers. In spring, concentrate on hive expansion, swarm prevention, and increasing brood production. Summer necessitates attention to honey production and pest control. Harvesting honey and preparing the hive for winter by providing adequate food storage occurs in the fall.

In the winter, keep hives warm and watch for symptoms of hunger or moisture buildup. Long-term success in beekeeping depends on adapting management approaches to the colony's seasonal needs.

CHAPTER FOUR
Knowing When to Harvest

Assessing Honey Ripeness Before harvesting honey, it is critical to assess its ripeness. Look for capped cells and determine their moisture content. A moisture content of approximately 18% is optimal. Turn the frame upside down and shake to see if the honey remains in place; if so, it is ready. Also, pay attention to how the color and aroma change.

instruments & Equipment Required for Honey Extraction Having the appropriate instruments makes honey extraction easier. A honey extractor, uncapping knife or fork, bee brush, and strainers or filters are all required. Also, keep jars, lids, and a clean workspace handy. These tools ensure that the extraction process runs smoothly and efficiently.

Extractors, Filters, and Uncapping Techniques When extracting honey, it is necessary to uncap it carefully.

To remove wax caps from frames, use a fork or uncapping knife. Then, place the frames in a honey extractor and spin to release the honey. Filtering removes dirt and ensures purity. There are several uncapping processes, including hot knife and cold knife approaches, each having advantages.

Bottling and Storing Honey: Hygiene Practices It is critical to maintain hygiene when bottling and storing honey. Use clean, sterilized jars and lids.

Fill jars with minimal air space to avoid crystallization. Honey should be stored in a cold, dark place to retain its flavor and quality. To avoid contamination, clean your equipment on a regular schedule.

Tips for Maximizing Honey Yield There are various aspects to consider while maximizing honey yield. Provide ample floral sources near the hive for bees to forage. Regular hive inspections assist in determining the best time to harvest. Maintaining healthy hives requires sufficient diet and pest management. Furthermore, make sure the hive has enough space to store honey. These measures improve honey output and quality.

Consider generating alternative honey products such as comb honey, creamed honey, and pollen. Comb honey provides a natural, unprocessed solution. Creamed honey, created through controlled crystallization, has a smooth texture and is easy to spread. Collecting and selling pollen can also help you earn money while improving your health. Explore these possibilities to vary your honey business and accommodate a wide range of consumer tastes.

CHAPTER FIVE
Importance of Providing Adequate Nutrition to Bees

Bees, like other living organisms, require a sufficient diet to stay healthy and productive. Adequate nutrition ensures that bees have enough energy and resources to forage, produce honey, and rear brood successfully.

A bee's diet must include pollen, nectar, and water, which provide them with the carbohydrates, proteins, vitamins, and minerals they require for survival.

Without sufficient nourishment, bees can become weak, susceptible to infections, and less efficient at their activities, affecting the overall health of the colony.

Understanding Pollination: The Importance of Bees and Crops

Pollination is an important process for bees and crops, resulting in a symbiotic connection. Bees take nectar and pollen from flowers while foraging, inadvertently moving pollen grains from one flower to another, aiding fertilization.

This technique is critical for the reproduction of flowering plants, including many crops that rely on bee pollination to produce fruits and seeds.

Pollination provides bees with a consistent food source and adds to the diversity and richness of plant species in their surroundings, ultimately benefiting bee populations and ecosystems.

Supplemental Feeding: Sugar Syrup and Pollen Patties

Supplemental feeding is a technique used by beekeepers to supply extra nutrition to bee colonies when natural supplies are limited or unavailable. Sugar syrup, formed by dissolving sugar in water, replaces nectar when flowers are not flowering or providing enough nectar. Pollen patties, created from a combination of pollen, sugar, and other nutrients, replenish the bees' diet with important proteins and vitamins, especially amid pollen shortages. These feeding practices enable bee colonies to survive lean times and boost their development and productivity all year.

Natural Pest and Disease Control Methods.

Maintaining bee health necessitates effective pest and disease management measures that reduce the use of chemicals while encouraging natural resistance within the colony. Regular hive inspections, adequate cleanliness, and habitat management all contribute to a lower chance of pest infestations and disease outbreaks.

Furthermore, strategies such as integrated pest management (IPM), which integrates cultural, biological, and mechanical control measures, provide long-term solutions to pest and disease control while reducing harm to bees and the environment. Beekeepers may keep their colonies healthy and thriving by putting prevention first and using natural control measures.

CHAPTER SIX

Recognizing and treating common bee ailments

Beekeepers must be diligent in identifying common diseases that can harm bee colonies and taking fast action to treat them. Diseases including American foulbrood, European foulbrood, and chalkbrood, as well as pests like Varroa mites, can weaken or wipe out bee colonies if not handled.

Regular hive inspections enable beekeepers to spot indicators of disease or infestation, including aberrant brood patterns, strange behavior, or visible parasites.

Treatment options vary depending on the ailment but may involve cultural traditions, biological controls, or the administration of licensed pharmaceuticals. Timely action is critical for limiting disease transmission and maintaining the health of bee colonies.

Integrating Medicine: When and How to Use Them

While natural pest and disease control measures are favored, pharmaceuticals may be required to address major threats to bee health. Beekeepers should carefully analyze the severity of the pest or illness, taking into account colony strength, environmental circumstances, and potential threats to adjacent colonies or nearby wildlife. When drugs are judged required, they should be used by label directions and local legislation to guarantee efficacy while minimizing harmful effects on bees and the environment.

Using drugs sparingly and only as a last resort will help reduce hazards while protecting bee colonies' health and productivity.

Evaluate Hive Strength for Expansion.

To successfully increase your apiary, first examine the strength of your existing hives. Look for evidence of a growing colony, such as a healthy brood pattern, adequate honey reserves, and a large number of worker bees. A hive with these traits is ready for expansion, ensuring that new colonies have a strong foundation to thrive.

Splitting Hives: Advantages and Procedures

Splitting hives is an important method for growing your apiary while protecting the health of your current colonies. The method is separating a strong hive into two or more distinct colonies, each with its own queen and resources. This increases the number of hives while simultaneously preventing overpopulation and swarming, resulting in a more profitable beekeeping enterprise.

Purchasing Additional Colonies

When you need to grow your apiary rapidly, obtaining more colonies can be a practical choice. Look for trusted providers who sell healthy bees and guarantee they come from disease-free areas.

Introducing new colonies into your apiary can increase productivity and diversity, allowing you to meet your beekeeping goals more efficiently.

CHAPTER SEVEN

Queen Rearing: Methods for Sustainable Beekeeping

Mastering queen-rearing strategies is critical for long-term success in beekeeping.

Selectively breeding queens with desirable features such as docility, illness resistance, and honey output will help your colonies' overall health and productivity. Whether through grafting, queen cell modification, or other ways, investing in queen-rearing abilities enables beekeepers to maintain healthy, robust bee populations.

Managing the Transition to Commercial Beekeeping

The transition from amateur to commercial beekeeping involves significant preparation and study. Assess market demand, production capacity, and financial resources to see if commercial beekeeping is a realistic option for you. Create a business strategy outlining goals, strategies, and potential problems, and seek advice from experienced beekeepers or industry experts to effectively negotiate this huge transformation.

Scaling Up: Infrastructure and Business Consideration

As you expand your beekeeping enterprise, invest in infrastructure and business concerns to ensure growth and efficiency. Upgrade equipment, such as hive boxes, extractors, and protective clothing, to accommodate increased honey production. In addition, build efficient methods for honey extraction, packaging, and distribution to successfully satisfy market demands. To achieve long-term success in commercial beekeeping, prioritize sustainability, scalability, and profitability.

Branding and Packaging: Creating an Attractive Product Your honey's packaging and branding are critical to attracting customers.

Create labels and packaging that highlight the quality and originality of your honey. Stand out on the shelf with clear, succinct language and eye-catching imagery. Consider eco-friendly packaging solutions to attract ecologically conscious customers.

Identifying Target Markets: Local or Regional, Understanding your target market is essential for effectively selling your honey products. Determine whether your concentration will be local, regional, or Internet. Local markets may comprise nearby communities and farmers' markets, whereas regional markets may encompass neighboring towns or cities.

Pricing Strategies: criteria to Consider Setting the appropriate pricing for your honey products necessitates careful evaluation of a variety of criteria. Consider the cost of manufacturing, which includes beekeeping equipment, labor, and packaging supplies. Analyze market demand and competition to find a competitive yet profitable price point. To ensure long-term viability, consider overhead costs and profit margins.

CHAPTER EIGHT

Building Customer Relationships: Education and Engagement Establish great relationships with your consumers by offering educational and engagement opportunities. Provide instructional sessions or workshops to educate consumers on the benefits of honey and beekeeping techniques. By educating and engaging customers, you may build trust and loyalty, resulting in repeat business and referrals.

Treating Bee Stings: First Aid and Prevention

First, if stung, be cool and quickly remove the stinger to avoid further venom. Use a cold compress to alleviate discomfort and swelling. Wear protective gear, be cautious around hives, and avoid floral scents. Keep an epinephrine injector available for severe allergic reactions, and seek medical attention right away if necessary.

Addressing Neighborhood Concerns: Education and Communication

Communicate openly with your neighbors about the benefits of bees and address any anxieties or misconceptions they may have. Offer to distribute honey or give a tour of your hives to explain their importance.

Assure them that you are making efforts to protect their safety, such as keeping hives away from property lines and providing appropriate water sources for bees.

Managing Aggressive Bees: Tips to Calm Them

To avoid provoking a protective reaction, maintain a cool approach when among bees. Use smoke to quiet bees during hive inspections, and wear protective clothes to avoid stings. If you meet angry bees, walk away slowly and gently, without swatting or making sudden movements. Consider requiring aggressive colonies to improve their behavior.

Legal Issues and Regulations: Permits and Liability Insurance.

Before starting a beekeeping operation, research local regulations and secure any necessary permits. Invest in liability insurance to cover yourself in the event of an accident or property damage involving your bees.

Ensure that you are by local zoning laws, hive-site standards, and any other beekeeping-specific regulations.

CHAPTER NINE

Managing Swarm Calls: Protocols and Safety Measures

Establish standards for reacting to swarm calls, such as analyzing the swarm's position and accessibility. Use a bee vacuum to safely catch the swarm, or gently shake it into a hive box. Wear protective clothing and secure the area to keep onlookers from being stung. After that, move the swarm to an appropriate area, ensuring that they have access to food and water.

Identifying and Addressing Queen Issues:

If your hive behaves abnormally, it could be due to queen-related concerns. Supersedure, in which bees replace an aged or failing queen, is a natural phenomenon. Absconding, on the other hand, occurs when the entire colony flees the hive, typically owing to stress or disease. Queenlessness occurs when a hive loses its queen and does not have a replacement. To address these difficulties, inspect your hive regularly for signs of a failing queen, keep an eye out for absconding behavior, and be prepared to replace the queen if necessary.

Coping with Robbery Behavior: Prevention and Intervention

Robbing in beehives can be annoying and harmful. To avoid it, keep your hives well-ventilated and avoid spills or leaks of honey or sugar syrup, which can attract robbers. If you see robbing, close the hive entrances to make it easier for guard bees to defend the colony.

Consider installing entrance reducers or screens to dissuade burglars. Intervention may include temporarily shutting the hive entrance, placing diversion feeders away from the afflicted colony, or relocating weaker colonies to reduce the danger of robbery.

Managing Varroa Mites: An Integrated Pest Management Approach

Varroa mites pose a severe threat to bee colonies, but they can be effectively managed using integrated pest management (IPM) measures. These include using screened bottom boards to check mite levels, drone brood traps to control mite numbers, and organic therapies like formic acid or thymol. Regular monitoring is essential for detecting mite infestations early, and alternating treatment approaches can help prevent mites from becoming resistant.

Recognize and Prevent Colony Collapse Disorder (CCD).

Colony Collapse Disorder (CCD) occurs when the majority of worker bees quit the hive, leaving behind the queen and brood. While the actual origin of CCD remains unknown, contributing factors could include pesticide exposure, infections, and stresses such as low nutrition or habitat loss. To avoid CCD, keep a diversified floral landscape around your apiary, avoid using pesticides near hives, and check hive health routinely. Respond quickly to any indicators of stress or sickness to help reduce the risk of CCD.

CHAPTER NINE

Handling Wax Moths, Small Hive Beetles, and Other Pests.

Wax moths and small hive beetles are typical pests that can cause havoc in beehives if left unchecked. To control these pests, keep your colonies strong and healthy, and inspect your hives regularly for symptoms of infestation, such as webbing or larvae.

Physical controls, such as freezing or sun-heating infected hive components, can assist in eliminating wax moth larvae and eggs, whilst beetle traps or oil traps can help manage small hive beetles. Maintaining proper hive hygiene, such as cleaning debris and excess wax, can also help to deter pest infestation.

Joining Beekeeping Associations: Benefits and Opportunities.

Beekeeping clubs provide numerous benefits to novices. Joining one gives you access to a supportive network of other beekeepers who may offer advice, share their experiences, and provide resources. Associations frequently host activities including meetings, workshops, and field trips where you can learn new techniques, share expertise, and network with experienced beekeepers.

Furthermore, many societies offer discounts on beekeeping materials and equipment through collective purchase arrangements, allowing you to save money as you begin your beekeeping journey.

Taking part in workshops, conferences, and training programs.

Beginners must attend workshops, conferences, and training programs to get practical beekeeping skills and knowledge. These events provide hands-on learning opportunities in areas like hive care, pest control, honey extraction, and more. Expert educators will walk you through the foundations of beekeeping, delivering useful insights and techniques that you can apply to your own apiary. Through engaging sessions and demonstrations, you will build confidence and expertise in caring for your bees.

Building a Beekeeping Library: Recommended Books and Publications.

Building a comprehensive beekeeping library is vital for beginners to gain a better understanding of beekeeping methods and skills. Investing in suggested books and magazines offers you authoritative materials provided by seasoned beekeepers and industry professionals. These books discuss a variety of issues, including bee biology, hive management, illness control, and honey production. With access to reputable reference resources, you may gradually improve your knowledge base, troubleshoot problems, and fine-tune your beekeeping skills.

CHAPTER TEN

Mentorship and Networking: Learn from Experienced Beekeepers

Seeking mentorship and networking opportunities with experienced beekeepers is beneficial for novices who want to accelerate their learning curve. Mentors offer specialized counsel, advice, and support based on your specific needs and circumstances.

Shadowing expert beekeepers and witnessing their practices firsthand can provide you with practical insights into beekeeping techniques, hive management strategies, and problem-solving approaches. Networking with other beekeepers also leads to new learning opportunities, collaborations, and friendships within the beekeeping community, which enriches your beekeeping journey.

Finally, starting out in bee farming, also known as apiculture, is a potential endeavor with numerous rewards. Through this path, novices can not only help the preservation of critical ecosystems but also reap a variety of economic and personal benefits.

First and foremost, bee farming provides a unique opportunity for newcomers to directly engage with nature.

Individuals who care for bee colonies contribute to the important role of pollination, promoting the growth of different flora and ensuring agricultural productivity. This symbiotic link with the environment promotes a better awareness and appreciation for ecological interconnectedness.

Second, from a financial standpoint, beekeeping offers significant opportunities for beginners. While early investments in equipment and education are required, the relatively cheap overhead expenses and small space needs make it a viable entrepreneurial venture. With proper care and management, beekeepers can make money by selling honey, beeswax, pollen, royal jelly, and even pollination services to local farmers.

Furthermore, beyond its economic significance, bee farming provides newcomers with a rewarding journey of personal development and fulfillment. The rigorous attention and care necessary to tend bee colonies instill a sense of responsibility and attentiveness. Beginners learn patience, resilience, and a deep appreciation for nature's resiliency as they monitor the intricate dynamics within the hive and see the results of their labor in the form of honey production.

Furthermore, bee farming facilitates community engagement and knowledge exchange. Beginners can interact with other beekeepers via local groups and workshops, forming a supportive network for learning and collaboration. Beginners can better handle hurdles and contribute to the growth of apiculture techniques by sharing their experiences and insights.

In essence, bee farming for beginners is more than a pastime or a business; it is a voyage of discovery, stewardship, and enrichment. As newcomers embark on this journey, they plant the seeds of a sustainable future for both themselves and the environment. With dedication, passion, and a willingness to learn, beekeeping beginners can embark on a meaningful and rewarding path that goes beyond honey production to enrich their lives and the world around them.

www.ingramcontent.com/pod-product-compliance
Lightning Source LLC
Chambersburg PA
CBHW072018230526
45479CB00008B/283